Foreword

Lack of safe and reliable water supply affects more than 800 million people worldwide. Globally, this has a major negative impact on human health, environmental sustainability and on economic development. Rapid population growth, urbanisation, climate change, pollution and inadequate financing present unprecedented challenges to the provision of water and sanitation services.

Improving and managing universal services of water and sanitation in a holistic manner is critical to achieving the United Nations Sustainable Development Goals, and addressing the needs of millions of people around the world. Ensuring access to water services is a key factor in working towards the Goals, and water service delivery planning can support utilities in improving provision of these services. A service delivery plan identifies the actions required and associated costs for achieving a defined level of water services delivery over a defined period.

Over the years, the International Water Association (IWA) has developed several frameworks and guidance for utilities to improve water services. These include the IWA Bonn Charter for Safe Drinking Water which established a framework for the collective implementation of integrated risk assessment and management systems aimed at ensuring the safe management of drinking water. This provided a foundation for developing the IWA/WHO Water Safety Planning Manual, which has played an important role in improving the provision of safe drinking water.

However, there was a need to support utilities with limited resources by providing guidance on the implementation of water services now and in the future with changing populations, climate, land use and urbanisation. Utilities are usually required to develop strategic plans by their regulators; however, service delivery plans are not mandatory. Yet they are extremely useful as they can provide actionable steps to implement strategic plans including the measures required and associated costs for achieving a defined level of water services delivery over a defined period.

This document is a guide or how-to manual on preparing water service delivery plans with a focus on small to medium-sized water utilities having approximately 5,000 to more than 100,000 connections mainly in areas with limited capacity and resources. The manual is simplified enough to ensure that these utilities are able to move from a situation where they might be struggling to deliver water services to one in which basic service levels in terms of water quality, quantity, accessibility, reliability, affordability and acceptability are met. Meeting these basic service levels provides a strong foundation for the utility to progressively move up the ladder of delivering improved services.

Kalanithy Vairavamoorthy, Phd
Executive Director, International Water Association

Ibrahim Rifath

Guidance on Preparing Water Service Delivery Plans

A manual for small to medium-sized water utilities in Africa and similar settings

Kizito Masinde *International Water Association*
Michael Rouse *University of Oxford*
Martha Jepkirui *Namano Mibey Company Limited*
Katharine Cross *International Water Association*

Published by   **IWA Publishing**
                          Republic – Export Building, 1st Floor
                          2 Clove Crescent
                          London E14 2BE, UK
                          Telephone: +44 (0)20 7654 5500
                          Fax: +44 (0)20 7654 5555
                          Email: publications@iwap.co.uk
                          Web: www.iwapublishing.com

First published 2021
© 2021 IWA Publishing

British Library Cataloguing in Publication Data
A CIP catalogue record for this book is available from the British Library

ISBN: 9781789062434 (print)
ISBN: 9781789062441 (eBook)

DOI: 10.2166/9781789062441

This document is an output from the REACH programme funded by UK Aid from the UK Foreign, Commonwealth and Development Office (FCDO) for the benefit of developing countries (Programme Code 201880). However, the views expressed and information contained in it are not necessarily those of or endorsed by FCDO, which can accept no responsibility for such views or information or for any reliance placed on them.

Content

Acknowledgements

This publication, Guidance on Preparing Water Service Delivery Plans, is the outcome of the International Water Association's partnership as part of REACH (https://reachwater.org.uk/), a global research programme to improve water security for the poor by delivering world-class science that transforms policy and practice. The seven-year programme (2015–2022) is led by Oxford University working with an international consortium of partners and funded with aid from the UK Government.

We thank all those who contributed to the preparation of this how-to manual for their invaluable inputs and insights. We also acknowledge the following individuals and organisations for their contribution in reviewing it:

Adesola Adedugbe (Independent Consultant), Alan Wyatt (Independent Consultant), Augustin Boer, BDO Business Advisory SRL, Bambos Charalambous, Hydrocontrol Ltd, Demissie Eskendir Alemseged, African Development Bank, Dennis D. Mwanza, RTI International, Francis Umemezia, AfWA focal point in Nigeria, Gustaff Chikasema, Lilongwe water board in Malawi, Inês Breda, Silhorko-Eurowater A/S, Joe Dalton, Freelance Consultant, Lisa Broß, Wasserversorgung Rheinhessen-Pfalz GmbH, Rob Hope, University of Oxford, Rory Moses McKeown, World Health Organization, and Rui Sancho, Águas do Algarve.

Finally, we express our gratitude to the designers, editors and those who have provided overall leadership in the completion of this guide.

The findings, interpretations and conclusions expressed herein are those of the authors and do not necessarily reflect the views of the IWA.

©Magdalena Paluchowska

1. **Introduction**

This manual is a guide on how to prepare a service delivery plan for small to medium-sized water utilities (supplying approximately 5,000 to more than 100,000 customers) in Africa and similar settings. It is intended for utilities operating in limited resource settings, and/or working with limited data and information. It is a useful reference for utilities embarking on developing their first water service delivery plan, and for those intending to improve their existing service delivery planning. This type of plan identifies the actions required and associated costs for achieving a defined level of water services delivery over a defined period.

The manual is in two parts:

Part 1: Explains how to use cost and financial models to produce a new delivery plan.

Part 2: Provides utilities with the tools used to develop the service delivery plan for the fictional Milele Town in Kenya, so that utilities can adjust and develop it to suit their own context.

A service delivery plan consists of three components:

1. A narrative on the current situation, the various plans, and programmes to be pursued, and a summary of the costs.

2. **A spreadsheet-based cost plan** which gives the cost assumptions, presents the current operating costs and how they will change during the period of the plan, and the costs of planned developments for a utility. A development is a set of activities (e.g. improved revenue collection) that are required for the utility to achieve the desired level of service. The bottom line provides total annual costs for the planning period and required tariff calculations.

3. **A spreadsheet-based financial model** which connects with the cost plan and can make comparisons between funding options.

Note: The cost plan and the financial model are integrated in the one spreadsheet-based system. The relationship of the contents of a business plan, a master plan and a strategic plan can be seen in the diagram in Annex 1.

Utilities are normally required to develop business plans, master plans or strategic plans by their regulators. Business plans seem to be most commonly based on the number of guiding documents available; for example, the Water Services Regulatory Board (WASREB) Guideline on Business Planning 2019[1] and the UN Habitat's Guidelines for Preparing a Business Plan for Urban Water Utility[2]. As illustrated in Annex 1, there are some differences and commonalities between the three types of plan.

A master plan is at a high level and has a long-range vision on direction and development. A strategic plan is for a shorter period (3–5 years) and is used for implementing and managing strategic direction. A business plan details goals and how these will be achieved by whom and in what timeframe.

A service delivery plan forms the bulk of a business plan but takes it to the next step by converting a business plan into real actions, showing what is important in the short, medium and long term. A service delivery plan helps utilities to think beyond infrastructure but also to consider non-technical issues such as training which are equally important for achieving their goals. Service delivery plans are in most cases not mandatory, and guidelines for developing them are not provided. This manual is important in covering this gap and ensuring that utilities know how to move forwards once they have developed their strategic plan.

It is important to note that some essential exercises necessary in water service delivery, such as financial forecasts, equity in service delivery and determination of non-revenue water, are outside the scope of this manual. Users are advised to seek help for instruction in fulfilling such needs. The manual covers the basics of service delivery planning to ensure that small utilities can move themselves from an 'unstable' to a 'stable' position, after which additional planning activities can be introduced. Assumptions made in the cost and financial plans are specific for the example given. Users will need to consider their local situation when making assumptions.

Water service delivery can help attain Sustainable Development Goal 6, specifically target 6.1: by 2030, achieve universal and equitable access to safe and affordable drinking water for all. If a water service provider includes Sustainable Development Goal 6.1 as a main goal, or is instructed to do so by its government, the universal coverage element provides for equity of water supply by 2030, including provision for those in poverty. In the shorter term, additional kiosk-type provision can be made for those living within reach of the distribution system, and for other temporary approaches, such as tanker-fed community-managed systems, to be put in place, until the distribution system has been extended to achieve universal coverage. In parallel, water service delivery responds to the UN Resolution of Water and Sanitation as a Human Right, which gives specific focus to equity of service in relation to poverty, including that water should be affordable as well as safe and accessible. The UN Resolution also addresses sustainability, with loss of a service being related to loss of a human right.

Assumptions on available water resources have to be made on the basis of best estimates of population (usually growth), the best knowledge available on rainfall predictions and on storage provisions, whether reservoir capacity or groundwater recharge. Climate change has increased the frequency of extreme rainfall events, both wet and dry; consequently, plans must be based on the likelihood of periods (years) of either dry or wet conditions. If an area is currently receiving greater than average rainfall, plans cannot be made on the continuation of 'wetter' conditions but must take into account future dry periods [3].

Two excel files are attached to this document. To access these files click on the paperclip symbol on the sidebar in Acrobat Reader where they can be accessed. The files include:

1. 2020 Milele Town Services Delivery Cost Plan *(example of a spreadsheet based cost plan)*

2. MIWASCO General Water Supply financial model *(example of a spreadsheet-based financial model)*

1.1. Service delivery planning

Figure 1. Steps to develop a service delivery plan

To bring improvements in service delivery, a delivery plan can be developed by following five simple steps (Figure 1).

1. Determine the current situation and status of performance indicators

The current situation can be assessed using available performance indicators. In the case of Kenya, water utilities can use the key performance indicators (KPIs) that are reported to the regulator, WASREB, as data already exists. These include water coverage; drinking water quality; hours of supply (hours per day); non-revenue water; metering ratio; staff productivity (staff per 1,000 connections); personnel expenditure (as a percentage of operation and maintenance costs); revenue collection efficiency; operation and maintenance cost coverage; sewered sanitation coverage; and sanitation coverage.

However, water utilities should not limit themselves to set indicators from a regulator that are used for reporting. They can include indicators that are best suited to match their strategic plans. This step also involves identifying and filling in any data gaps. The areas with data gaps or no information should be noted as they will inform actions for the first year of the delivery plan in step 3.

2. Set targets for the performance indicators

The water utility at this stage must determine what it aims to achieve in the 5-year delivery period with achievable targets. This step, together with steps 3 and 4, is performed iteratively so the utility has sufficient revenues and financing in the 5-year period for these targets.

3. Determine the actions needed to meet targets

For each action determined, the who (who will be implementing), the when (when it will be implemented) and the cost should also be identified. Several actions may be required to address a single target. Approaching a KPI through multiple and coherent actions should be the goal.

4. Develop a cost plan

The cost plan brings together the utility's annual operation and maintenance costs and the capital requirements for actions identified in step 3.

5. Develop the financial model

The financial model provides information that can be used in identifying current and future investment needs and options. This can be put together in a financial plan.

More details of the step-by-step process are given in the sections below.

2. Determining the current status of performance indicators

Establishing current service levels (step 1) is the baseline against which targets for the service provider will be determined. Such service levels may, for example, cover the water quality, quantity, accessibility, reliability, affordability and acceptability. The analysis includes interrogation of the current performance in the various components and KPIs compared with the ideal status as described in an organisation's strategic plan. In this manual, it is assumed that the utility will already have a strategic plan, or equivalent.

It is important to note that a strategic plan is different from a service delivery plan: the former defines the 'what' while the later defines the 'how'. The strategic plan outlines the mission, vision and values of the organisation, identifying the broad challenges to be solved and setting the guiding policies. The strategic plan is important as it informs decision making on how to allocate limited resources among core business functions and how to make investments that increase value for utility customers and stakeholders within acceptable levels of risk.

The service delivery plan determines the current status of KPIs and defines what steps to take to improve the KPI score with the overall aim of solving the challenges identified in the strategic plan. The service delivery plan determines the costs of the steps to be taken and defines responsibilities.

For a public utility, the key value for customers and stakeholders is usually accessing safe drinking water and sanitation services (if these are also provided) at affordable cost. If well developed, the goals will ideally be shaped by the local environmental, regulatory, political and economic considerations, as well as by the community and social agenda.

Several tools exist that can establish current service delivery performance; for example, AquaRating (https://aquarating.org/) developed by the International Water Association and the InterAmerican Development Bank, and the World Bank's Utility Turnaround Framework.

In the example given later in this manual for the Milele Water and Sanitation Company (MIWASCO), a fictional water utility that serves the fictional Milele Town, the current service delivery performance was determined on the basis of the WASREB-recommended performance indicators. Additionally, a cost plan and a financial model were used to establish the utility baseline costs and the costs of the necessary service delivery developments, as shown later.

3. Setting objectives/targets and identifying actions

Setting the targets for service delivery over the planning period (step 2) goes hand in hand with identifying the actions to achieve them (step 3). Service delivery plans are designed to convert the targets (the what) into several stages for delivery improvements (the how). Each stage will depend on priorities related both to current performance and to national and local priorities. These will vary between utilities.

This prioritisation of improvements in service then leads to the development of actions required to achieve specific targets for key KPIs. In the case of the plan for Milele Town, the following stages were followed:

Stage 1. Determine current basic operation costs and how they will change over the plan period.

Stages 2 and 3.
(a) Development 1: improved revenue collection.
(b) Development 2: increased supply coverage by network extension.
(c) Development 3: drinking water quality.
(d) Development 4: leakage reduction in the distribution system.
(e) Development 5: improving water security in relation to predicted dry periods.

Stage 4. Integrate stage 1 and all the developments into one plan to produce total annual costs over the period of the plan. The result includes a narrative text describing all the actions and their costs, and a cost plan in a set of spreadsheets.

Stage 5. Using the financial model to consider the funding options, with a combination of calculated tariffs, grants and loans.

4. Developing a cost plan

Developing a cost plan (step 4) involves costing target actions as well as designing an implementation plan. This cost plan will form a key input into the financial model. The details of the costing need to be worked out, usually in a separate file using data and information collected by the utility. This will include, for example, detailed costs of training staff, customer engagement through public meetings, household metering, installation of automated meters for kiosks, pumping, mapping of the distribution network, leak detection and leak repairs, non-revenue water (NRW) surveys, etc. An example of a cost plan is available as an attached file to this document, and details of the components within the cost plan are in part 2 of this manual.

The type of data and information required to compute costs will be determined from the objectives that are set, and actions that are identified in step 3. An example of the data and information that were obtained from MIWASCO to develop the cost plan is provided in section 5. A summary of the cost plan is contained in the example financial model, also described in section 5.

5. Using the financial model

Table 1. Proposed developments for MIWASCO as they appear in the Cost Plan

Guidance Document Developments	MIWASCO Developments	Activity item
		OPERATING COST ITEMS
Improved revenue collection	Increase revenue collection efficiency to 90%	Check and update customer database
		Debt collection
		Installation of 5 No. water ATMs in operational water kiosks
		Installation of improved customer and meter reading database
		Installation of standardised meters
		Capacity building for staff
		Customer engagement through regular public meetings
Increased supply coverage by network extension	Network expansion of the distribution system	Expansion of the network by an average of 10 km p.a
		Detailed mapping of distribution network
Distribution system leakage reduction	Leakage reduction from 66% to 30%	Metering of all production boreholes
		Set up of leak detection teams with training and capacity building
		Active leak detection and leak repairs
		Purchase of leak detection equipment
		Stock of materials for leak repairs
		Creating a network model, Conduct comprehensive NRW survey to determine physical and commercial losses and design of DMAs
		Replcement of PVC pipeline with HDPE
Drinking water quality	Improving drinking water quality	Train staff on water saftery planning
		Prepare first water safety plan
		Increase storage
		Construction of water kiosks

To use the financial model (step 5), various data and information from the utility are required. The example financial model has a spreadsheet-based file with four sheets:

1. Cost plan sheet
2. Assumptions sheet
3. Inputs sheet
4. Summary statements sheet

5.1. Populating the spreadsheets

The spreadsheets in the example financial model have been used to create several developments which already have formulae built into them (see Table 1). The developments for the example financial model that are detailed in part 2 include the following:

- improved revenue collection;
- increased supply coverage by network extension;
- drinking water quality;
- distribution system leakage reduction.

The parameters in the formulae are set by the user. Once the required information is inserted into the 'Assumptions' and 'Inputs' worksheets, results are automatically worked out in the Summary Statements sheet as per the pre-set formulae. It is therefore important that accurate information is obtained and filled into the 'Assumptions' and 'Inputs' worksheets. Data should be entered into the yellow-coloured cells.

5.1.1. Basic assumptions, water tariff and population

Populating the financial model begins with the Assumptions sheet (see sections of the Assumption sheet in Tables 2 and 3). Information on (1) basic assumptions, (2) tariff charges and (3) population figures needs to be available. The user should have access to the information; if it is not available, the user will need to make estimates as well as take steps to obtain the data.

1) Basic Assumptions

(a) Start Year: the year in which data on inputs into the worksheet have started.

(b) Average Household Size: the average number of people within each household in the service area. This is obtained from census data.

(c) Other Expenses as % of Revenue: other costs apart from salaries, chemicals, electricity and new connection costs, expressed as a percentage of the total revenue of the utility. This can be obtained from the utility financial records.

(d) Other Current Assets as % of Revenue: this is income from other sources apart from water consumption and sewerage revenues, for example income from a training centre, expressed as a percentage of the total revenue.

(e) Long-Term Investments as % of Rev.: the amount of revenue put into a long-term investment plan expressed as a percentage of the total revenue.

(f) Current Liabilities as % of Revenue: the amount of debt owed to others expressed as a percentage of the total revenue.

(g) Average Interest Rate on Debt: the estimated percentage interest charged on the utility debt.

(h) Average Interest Rate: LT Investments: the estimated interest rate obtained from the long-term investment.

(i) Average Tax Rate: this is tax charged on the utility yearly expressed as a percentage.

(j) Depreciation Rate: the estimated rate of depreciation of the utility assets.

Table 2. A view of the Basic Assumptions part of the Assumptions sheet

UTILITY FINANCIAL MODEL: ASSUMPTIONS			
Version	1.0		
Date	01/04/2020		
Currency	2018 KES		
Utility name			
Basic Assumptions			**Source**
Start Year			
Average Household Size			
Other Expenses as % of Revenue			
Other Current Assets as % of Revenue			
Long -Term Investments as % of Rev.			
Current Liabilities as % of Revenue			
Average Interest Rate on Debt			
Average Interest Rate: LT Investments			
Average Tax Rate			
Depreciation Rate			

Table 3. Assumptions sheet view of the Tariff and Population in Service Area section

Tariff											
Note: Simple volumetric tariff assumed											
	Current	2	3	4	5	6	7	8	9	10	11
Water: Household (per m3)											
Water: Industrial (per m3)											
Sewerage: Household (per m3 water)											
Sewerage: Industrial (per m3 water)											
Connection Fee: Household											
Connection Fee: Industrial											

Population in Service Area											
	Current	2	3	4	5	6	7	8	9	10	11
Total Population											
Estimated Households											

2) Tariff

Note: a simple volumetric tariff is assumed.

(a) Water: Household (per m^3): the amount of money the utility charges for each cubic metre of water consumed by a household.

(b) Water: Industrial (per m^3): the amount of money the utility charges for each cubic metre of water consumed by an industrial customer.

(c) Sewerage: Household (per m^3 water): the amount of money the utility charges for providing sewerage services for each cubic metre of water released by a household.

(d) Sewerage: Industrial (per m^3 water): the amount of money the utility charges for providing sewerage services for each cubic metre of water released by an industrial customer.

(e) Connection Fee: Household: the fee charged to a household for the authorised connection of a consumption point.

(f) Connection Fee: Industrial: the fee charged to an industrial customer for the authorised connection of a consumption point.

3) Population in Service Area

(a) Total Population: the total number of people living in the utility service area. This is obtained from census data.

(b) Estimated Households: the estimated number of households within the area served by the utility. This is obtained from census data.

5.1.2. Information requirements for the Inputs sheet

For the inputs sheet, the following datasets are required: (1) Water Production; (2) Water Consumption and Revenues; (3) Expenses; (4) Assets; (5) Liabilities and Equity; (6) Statement of Cash Flows.

(1) Water Production

(A) WATER PRODUCED

(i) Total Production Capacity (m^3): the volume of water the utility is designed to produce at its optimum. The information is obtained from the engineering designs of the treatment plant.

(ii) Total Production (m^3): the actual volume of water produced by the utility at its current state. The figure is obtained by measuring the amount of treated water that leaves the treatment plant.

(iii) Production as a % of Total Capacity: an expression of Total Production as compared with the Total Production Capacity (Water Treatment Plant Utilisation Capacity).

Note: for the MIWASCO example (detailed in part 2), as there was an absence of a centralised treatment plant, the production capacity of the boreholes was used.

(B) WATER AVAILABLE FOR CONSUMPTION

(i) Non-Revenue Water - Technical Losses (%): non-revenue water (NRW) is water that has been produced and is lost on the way before reaching the customer. It is also known as unaccounted for water. Technical Losses of NRW are losses due to leakage in the distribution network. More information on determining NRW can be obtained by studying the Standard IWA Water Balance.

(ii) Total Water Available (m^3): total water available is obtained by subtracting NRW from the Total Production.

(2) Water Consumption and Revenues

(A) HOUSEHOLD CUSTOMERS

(i) Total Connections: the total number of authorised household consumption points that a utility has on its record.

(ii) Consumption Per Connection with NRW (m^3): the average volume of water that a household consumes per year. It is obtained by determining the amount of water in the Total Production allocated to household supplies, divided by the Total Connections for households.

(iii) Consumption Per Connection without NRW (m^3): the actual average volume of water that a household consumes. It is obtained by determining the amount of water that is actually consumed by households, divided by the Total Connections for households.

- Water Sales: the total amount of money the utility makes through sales of water that is consumed by households (without NRW).

- Sewerage Fees: the total amount of money charged on households for the sewerage service provision.

(iv) Collection Efficiency: an expression of the amount of revenue collected compared with the total bill produced every month. It is expressed as a percentage.

(v) New Connections: the number of new authorised household consumption points that have been added to the Total Connections.

(B) INDUSTRIAL CUSTOMERS

(i) Total Connections: the total number of authorised industrial consumption points that a utility has on its record.

(ii) Consumption Per Connection (m^3): the average volume of water that an industrial customer consumes. It is obtained by determining the amount of water that is actually consumed by authorised industrial customers, divided by the Total Connections for industrial customers.

- Water Sales: The total amount of money the utility makes through sales of water that is consumed by industrial customers (without NRW).

- Sewerage Fees: total amount of money charged on industrial customers for the sewerage service provision.

(iii) New Connections: the number of new authorised industrial consumption points that have been added to the Total Connections.

(C) UNBILLED WATER

(i) Water Available for Consumption: total water available is obtained by subtracting NRW from the Total Production.

(ii) Total Consumed by Customers: the volume of water consumed by all customers. It is obtained by multiplying Total Connections by the Consumption per Connection. This is already worked out in the sheet.

(iii) Unbilled Water: the difference between Water Available for Consumption and Total Consumed by customers. This is already worked out in the sheet.

(iv) Unbilled Water as a % of Total Production: the percentage obtained by dividing the Unbilled Water by Total Production and expressing this as a percentage. This is already worked out in the sheet.

(D) MARKET SIZE

(i) Total Households in Service Area: the number of households in the area earmarked for service by a utility estimated from census data.

(ii) Total Actual + Potential Consumption (m^3): the total volume of water that is consumed with the estimated water that could be consumed if the remaining potential customers were connected. This is already worked out in the example sheet.

(iii) Total Value of Consumption: a compilation of the total potential earnings obtained by multiplying the water tariff by Total Actual + Potential Consumption (m^3) and adding to the result of multiplying the sewerage tariff by Total Actual + Potential Consumption (m^3). This is already worked out in the example sheet.

(3) Expenses

(A) FIXED COSTS

(i) Total Employees: the total number of staff employed by the utility

- Average Salary / Employee (Annual): the total salary expenditure divided by the Total Employees.

(ii) Depreciation: this is calculated from the depreciation rate and value of property, plant and equipment.

(iii) Other Expenses: this is calculated as a percentage of the revenue collected. This percentage is provided in the assumptions.

(B) VARIABLE COSTS

(i) Total Production (m^3): the volume of water the utility is designed to produce at its optimum. The information is obtained from the engineering designs of the treatment plant.

- Chemicals: Cost per m^3: the Total Production divided by the Total Cost of chemicals.

(ii) Chemicals: Total Cost: this is the cost of all chemicals used to obtain the total production.

- Electricity: Cost per m^3: the Total Production divided by the Total Cost of Electricity.

(iii) Electricity: Total Cost: the total cost of electricity used to obtain the Total Production.

- Repair and maintenance: Cost per m^3: the Total Production divided by the Total Cost of Repair and Maintenance.

(iv) Repair and maintenance: Total Cost: the amount of money spent on repair and maintenance in a year.

- % of Revenues Written Off / Year: the Bad Debt Expense expressed as a percentage of the total revenues.

(v) Bad Debt Expense: the average amount of debt that is written off every year.

(4) Assets

(A) CURRENT ASSETS

(i) Revenues

- % of Connections Late in Paying Bills: Household Customers: an estimation of the number of household customers who pay their bills late expressed as a percentage of the Total Household Connections.

- % of Connections Late in Paying Bills: Industrial Customers: an estimation of the number of household customers who pay their bills late expressed as a percentage of the Total Household Connections.

(B) NON-CURRENT ASSETS

(i) Land and Fixed Assets

- Land: Value of land owned by the utility.
- New Land Purchased: the value of new land acquired by the utility.

- Other Fixed Assets: value of other fixed assets such as buildings and equipment.

- Other Fixed Assets Purchased: value of Other Fixed Assets acquired in the year.

(ii) Property, Plant and Equipment: total value of land, fixed assets, other fixed assets. Historical values of this value are used and projections are made by estimating an appreciation rate.

Note: an annual appreciation rate of 9% was estimated for MIWASCO.

(5) Liabilities and Equity

(A) LIABILITIES

(i) Current Liabilities: the amount of money due to be paid to creditors within 12 months.

(ii) Debt: the total amount of money due to creditors of the utility.

(B) EQUITY

(i) Contributed Capital (Funding): funds contributed by the utility to acquire, upgrade, and maintain physical assets.

(ii) Donated Capital (Government Grants): funds donated by the government to acquire, upgrade, and maintain physical assets of the utility.

(6) Statement of Cash Flows

(A) INVESTING ACTIVITIES

(i) Capital Expenditures: funds contributed by the utility to acquire, upgrade and maintain physical assets.

Part 2

6. **The Milele Town service delivery plan, cost plan and financial model**

Table 4. Basic Assumptions made in the LOWASCO cost model

UTILITY FINANCIAL MODEL: ASSUMPTIONS

Version	1.0
Date	01/04/2020
Currency	2018 KES
Utility name	MIWASCO

Basic Assumptions		**Source**
Start Year	2015	
Average Household Size	7	*Based on WASREB LIA data (http://majidata.go.ke/public-portal/)*
Other Expenses as % of Revenue	30%	*Calculated from MIWASCO financial records*
Other Current Assets as % of Revenue	5%	*Assumed: MIWASCO records show a negative value for current assets*
Long -Term Investments as % of Rev.	3%	*Assumed*
Current Liabilities as % of Revenue	45%	*Calculated from MIWASCO financial records*
Average Interest Rate on Debt	10%	*Assumed*
Average Interest Rate: LT Investments	5%	*Assumed*
Average Tax Rate	30%	*Kenyan corporate tax rate*
Depreciation Rate	10%	*Rate used for plant, machinery and equipment*

Part 2 explains how the cost plan and financial model were created for each development in the MIWASCO case. As described in part 1, a development is a set of activities (e.g. improved revenue collection) that are required for the utility to achieve the desired level of service. It is meant to increase the understanding of how the financial model works and how to arrive at the figures in the spreadsheets.

To begin with, the following Basic Assumptions were made as shown in Table 4:

In the cost plan and financial model, it was assumed that the following costs would remain the same over the next 10 years:

(i) The simple volumetric tariff would stay constant at KES (Kenyan shillings) 33 per cubic metre for both households and industries.

(ii) The connection fee for households would stay the same at KES 1,200 and KES 5,000 for industries for the 10-year planning period.

The population in 2017 was used as the current population, obtained from the Kenya National Census figures. The population in subsequent years was estimated using the 3.35% population growth for Milele Town, also obtained from the Kenya National Census.

6.1. The cost plan

The cost plan aims to provide MIWASCO with an understanding of the current baseline operation and maintenance costs. These costs are important as they are the foundation upon which the other proposed developments are built. To develop the cost plan, the data and information below should be collected and made available. Presented data will be both historical (for at least the past 3 years) and projected, for the number of years the delivery plan is intended to cover, ideally at least 5 years and beyond.

Obtaining Total Operating Cost for Milele Town in the Financial Model

To obtain the total costs of providing services to Milele Town in a year, expenses outline in Table 5 need to be included.

Table 5. Expenses used to calculate total costs of providing water services in Milelet Town

COSTS	DESCRIPTION
SALARIES	Average annual salary per employee based on historical records and budgets
CHEMICALS	Annual volumetric cost calculated from annual expenditure on chemical and annual production
ELECTRICITY	Annual volumetric cost calculated from annual expenditure on electricity and annual production
REPAIR AND MAINTENANCE	Annual volumetric cost calculated from annual expenditure on repair and maintenance and annual production
NEW CONNECTIONS	Calculated from projected number of new connections and standard MIWASCO connection fee
BAD DEBT EXPENSE	Estimated on the basis of assumption that 25% of revenue is written off annually
DEPRECIATION	Calculated from accumulated depreciation and projected using an assumed annual depreciation rate of 10% for plant, machinery and equipment
OTHER EXPENSES	Covers all other expenses not included in the list and estimated from MIWASCO records to be approximately 30% of revenue annually

Note the following.

1. MIWASCO's financial year begins in July and ends in June. In the financial model, the years used should be interpreted as starting in July of the year. For instance, in the financial model, the year 2019 refers to the financial year 2019/2020, the year 2020 refers to the financial year 2020/2021.

2. The baseline as presented in the report only covers:

(a) the current costs;

(b) investment to improve basic data;

(c) Contributed capital of KES 1 million per annum and government grants of KES 2 million per annum; and

(d) debt reported in MIWASCO's 2016/2017 financial statements and assumed annual loan repayment of KES 200,000 on outstanding loans and an additional KES2,300,00 per annum for 4 years to clear debt to WRMA and WASREB.

3. The growth in the population to be served without investment to expand the distribution system.

4. The summation is already worked out in the 'Summary Statements' worksheet. For example, the Total Operating Cost for MIWASCO in 2019 can be found in cell G26 of the 'Summary Statements' worksheet.

5. In turn the information that goes into the summation in the 'Summary Statements' is obtained from the 'Inputs' worksheet; i.e. the information is first inserted as listed in section 5.1.2 (Information requirements for the Inputs Sheet).

6. Historical figures are actual figures obtained from MIWASCO records for the previous years; in this case from 2015 to 2017.

Projected figures are estimated figures of expenditure for each year in the future. They are computed to take into consideration planned expansions. These include both capital expenditure and operating expenditure.

6.2. Development 1. Improved revenue collection

In this development, the following costs were included.

6.2.1. Metering all customers

1. As of 2019, the number of active connections for MIWASCO was 8,251. The aim is that these connections grow by the following percentages of the previous year's total in the subsequent years. From 2024 onwards, the connections were calculated to grow at 5% above the previous year up to 2029.

2. The plan is to install standardised meters for all connections at the lump sum cost of KES 4,560,000. This total figure is arrived at by determining the number of meters required, and multiplying this cost by the cost of one meter.

6.2.2. Updating customer databases

1. Updating the customer database will include household and customer surveys to ensure that the following:

(a) all connections are correctly recorded in the database; i.e. customer details, location information, meter details;

(b) illegal connections are identified and handled accordingly.

2. With the existing zonal teams, the assumption was made that the teams would incorporate these surveys into their current operations. This measure is therefore expected to have a modest cost in its execution.

3. In the first year (2020), checking and updating the database is expected to cost more; i.e. KES 150,000. The exercise should cost up to KES 50,000 per year in the subsequent years. This is because what is required will be a routine update of the database with new customer information and changes to customer details, which are not expected to be many.

Table 6. Percentage used to calculate increase in connections

YEAR	2020	2021	2022	2023	2024
Percentage increase in connections	5%	10%	15%	15%	5%

6.2.3. Improvement to the current meter reading and billing system

1. MIWASCO currently uses two different types of software for meter reading and customer billing. Each month, meter readings are captured by the meter reading software, exported to .csv format and imported into Microsoft Access software for billing.

2. The cost plan provides for upgrading from the current separate system to one where both meter reading and customer billing functions can be performed within one platform.

3. It is estimated that the cost of obtaining this new software will be KES 3,000,000 in year 2.

6.2.4. Capacity building for staff

1. Capacity building for staff has been costed to cater for the setting up and training of a planning and monitoring unit.

2. It is estimated that this will cost MIWASCO KES 240,00 for the first and second years, and in the 6th and 7th years for the delivery plan.

3. It was assumed that the training in the first and second years would be sufficient to support the unit for the years in between when no training is planned.

6.2.5. Installation of five automated water dispensers in dormant water kiosks

1. Five kiosks are recommended to be fitted with automated water dispensers.

2. Each automated water dispensers is costed at KES 212,000, bringing the total to KES 1,060,000.

6.2.6. Customer engagement through regular public meetings

1. Customer engagement is identified as necessary to ensure that customers served within Milele Town are well informed on the water company's activities, the need to make timely payment of water bills and are aware of any upcoming or ongoing activities.

2. The costing has considered that customer engagement is a continuous exercise. The yearly cost of this is estimated at KES 630,000 for the 10 years of the delivery plan.

6.3. Development 2. Increased supply coverage by network extension

Milele's population over the next 10 years is estimated to increase by 3.35% annually. The population increase will take place both within and outside the current distribution network as the town expands.

6.3.1. Detailed mapping of distribution network

1. One objective of the proposed delivery plan is to establish a good understanding of the network and to achieve modelling capability.

2. In the early years much of the investment required is in distribution surveys to produce up-to-date network maps, and in capacity building of MIWASCO staff to perform network analysis.

3. The amount estimated for this is KES 4,000,000 in the first year. This is a lumpsum figure based on a comparison of previous mapping exercises.

6.3.2. Expansion of the network by an average of 10 km per annum

1. The investments costs estimated in this scenario cover the costs of network expansion and new connections in new coverage areas.

2. The assumption has been made that the water supply network of Milele will have to be extended by an average of 10 km every year to achieve full coverage of the service area in 10 years (see Table 6 for percentages used to calculate yearly network expansion).

3. The cost the 10 km network expansion every year is estimated to be KES 9,359,792. This is at the cost of KES 935,979.20 per kilometre.

6.4. Development 3. Drinking water quality

1. MIWASCO water quality analysis procedures are not documented, and reports/records of water quality assessments have not been provided.

2. There is a water quality team in place. MIWASCO reports to WASREB that they meet the drinking water quality standards as set by the Kenya Bureau of Standards.

3. Consumers in Milele use water from other sources in addition to the piped water.

4. This plan advises that the company develops a water safety plan (WSP) for its supplies and undertakes a safe drinking water and health campaign based on the WSP.

5. Developing the WSP is indicated as having no cost implication on the utility. This is usually the case where there is capacity already within that can train, develop and implement the WSP. Where this capacity is lacking, it is estimated to cost KES 500,000 in the first year.

6. Although a no-cost implication on the utility is assumed, when planning it is important for the utility to take note of ongoing costs for implementing the WSP, including time inputs by staff for reviewing and revising the WSP, WSP team meetings, and internal and external audits (see Table 7).

7. One training approach would be the following:

(a) training to be carried out annually for 3 years;

(b) in the second year, the training time required would be less as the WSP would have already be developed;

(c) the third year might cost the same as the first year to meet the need to audit WSP implementation and carry out a review;

(d) within 3 years, it might be expected that there would be enough capacity developed for trainers within the Company to carry out the training in-house;

(e) the drinking water health campaign could be incorporated into the customer engagement through regular public meetings already costed in scenario 2.

Table 7: Estimated cost of hosting an inhouse WSP training by MIWASCO

	ITEM	UNIT COST (KES)	NO. OF UNITS	NO. OF DAYS	TOTAL (KES)
i	Cost of trainer (5 days training, 1 day preparation, 2 days travel and 10 days support to develop the WSP)	10,000	1	18	180,000
ii	Travel for trainer	14,000	1	1	14,000
iii	Accommodation and subsistence for trainer	14,000	1	6	84,000
iv	One field trip	15,000	1	1	15,000
v	Printing of WSP training manual	3,000	20	1	60,000
vi	Printing of participants' workbook	500	20	1	10,000
vii	Conference charge for participants (including trainer)	1,000	21	5	105,000
	Total				468,000
	10% contingency				46,800
	Grand total				514,800

6.5. Development 4. Reduction in distribution system leakage

6.5.1. Metering of all production boreholes

1. Currently MIWASCO draws water from 12 boreholes. This plan recommends metering 10 of them.

2. The total metering cost is estimated at KES 500,000 per bulk meter.

6.5.2. Set up of leak detection teams with training and capacity building

1. To ensure there is capacity within MIWASCO to effectively detect leaks in the distribution system, there is need for training and capacity building on the same.

2. It is estimated that KES 350,000 will be required in the first year to train the team.

6.5.3. Active leak detection and leak repairs

1. The first year of the plan will focus mostly on purchasing equipment, setting up the team and training. It is therefore estimated that the amount of money required to carry out the leak detection and repair work will be KES 550,000.

2. From the second year to the tenth year, it is expected that the work of leak detection and carrying out repairs will increase. Each year has therefore been allocated KES 3,850,000.

6.5.4. Purchase of leak detection equipment

1. The amount of KES 90,000 has been allocated for the purchase of leak detection equipment in the first and second years.

2. It is expected that in the seventh and eighth years, there will be a need to replace and purchase more equipment. As such, the 2 years have also been allocated KES 90,000 each for this purpose.

6.5.5. Stock of materials for leak repairs

1. Leaks require urgent repair to save the utility money and reduce the possibility of water contamination in cases of intermittent water supplies. To avoid this, repair materials need to be stocked for ease of access after leak detection.

2. It is estimated that every year of the delivery plan will require KES 407,500 to stock these repair materials.

6.5.6. Creating a network model, conducting comprehensive NRW survey to determine physical and commercial losses and design of district metred areas

1. For effective control of NRW, the network model should be well understood. This will aid in determining the water balance and controlling of NRW, for example by designing district metred areas.

2. This exercise is expected to take place in the second year of the plan after the training. It is estimated to cost KES 2,000,000.

6.5.7. Replacement of PVC pipeline with HDPE

1. Owing to the advantages that have been identified for HDPE (high-density polyethylene) over PVC (polyvinyl chloride), particularly in Milele, one of the activities recommended is to replace PVC with HDPE.

2. This activity has been scheduled to begin in the fifth year and run to the ninth year of the plan at a cost of KES 24,500,000 every year.

6.6. Financing of the delivery plan

Funding options that are available for water supply services in Kenya are tariffs, grants and loans. Grants may be in the form of budgetary allocations, equalisation funds, decentralised funds, donor funded projects and development funds earmarked for water supply services.

Some of the options for financing of the development plan are as follows.

6.6.1. Tariffs

MIWASCO's current tariff of KES 33 per cubic metre is not sufficient to meet the baseline operating costs. The operating costs linked to the proposed developments will also not be recoverable using the current tariff. To achieve cost recovery, a tariff of KES 66 per cubic metre starting in year 1 would achieve cost recovery in year 4. To achieve immediate cost recovery, a tariff of KES 70 per cubic metre would be required. However, these two tariffs would not be feasible. Considering the economic status of most consumers in the region, the tariffs of KES 66 or 70 per cubic metre would not be affordable. An approval from the regulator for such a tariff would thus be unlikely. Even in the unlikely situation of regulator approval, the process of adjusting tariffs takes time; therefore the objective would not be achieved in the first year.

6.6.2. Combination funding

Combination funding is the option that has been used in the example financial model.

The options around funding are provided in the cost plan (see Table 8 below).

Firstly, option 2 means that the development plan would be financed by grants and a long-term loan for the extension of the distribution system. The option given here assumes KES 100 million is received as a long-term loan repayable over a period of 10 years at an interest rate of 14% [7]. The assumption is made that the loan would be disbursed in 10 equal instalments over that period.

In the financial model, this loan is included under debt in the Inputs sheet in year 1 of the development plan. The total amount of money repayable including interest is KES 174,391,399. This is added to the debt and begins to be repaid at the rate of KES: 17,439,140 per year. This amount is subtracted from the debt yearly onwards.

The cost of extending the distribution network accounts for 33% of the total development costs, which results in large grant requirements of approximately KES 19 million per annum. The financial model shows that, even with this funding option, the water utility will not be able to make a profit.

To reduce the amount of grant required, another option (option 3) was included where tariffs would be used to cover a third of the development costs to an average of KES 12 million per annum. The use of tariffs would reduce the amount of grant required in years 1, 3, 4 and 10 to less than that estimated in the baseline.

Table 8: Funding options for MIWASCO

KES	OPTION 1	OPTION 2	OPTION 3
Total funding requirements	281,243,186	281,243,186	281,243,186
Loan		93,597,927	93,597,927
Tariff	281,243,186		84,372,956
Grant		187,645,259	103,272,303

7. Annex

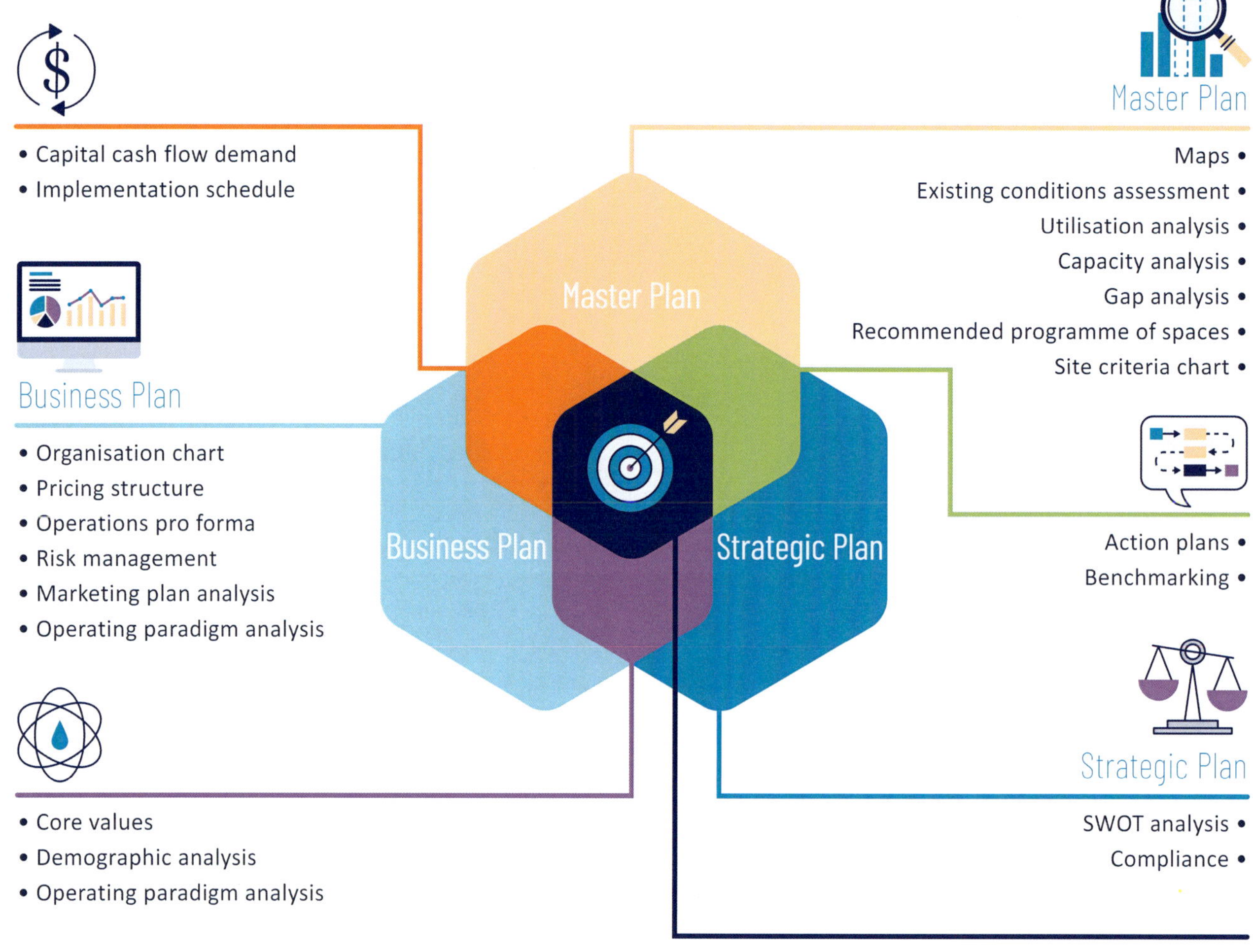

Figure 2. Contents of a business plan versus a master plan versus a strategic plan (adapted from a presentation by Drummie and Slater, 2013[6])

FOOTNOTES

[1] Water Services Regulatory Board (WASREB). (2019). Guideline on Business Planning. https://wasreb.go.ke/business-planning-guidelines/

[2] UNICEF. Guidelines for Preparing a Business Plan for Urban Water Utility https://mirror.unhabitat.org/downloads/docs/2527_1_595416.pdf

[3] Hall, J. W., Grey, D, Garrick, D., Brown, C., Dadson, S. J., and Sadoff, C. W. (2014). Coping with the curse of freshwater variability. Science, 24, 429-430.

[4] Water and Sanitation Program. (2008). Performance Improvement Planning Developing Effective Billing and Collection Practices. Available at: http://documents.worldbank.org/curated/en/713571468138288578/pdf/441190WSP0IN0P1ive0billing01PUBLIC1.pdf

[5] Water Services Regulatory Board. (2017). Software Requirements for A Model Billing System for The Utilities Technical Services Directorate WASREB/TSD/ER/1/2014-15. Available at https://wasreb.go.ke/water-service-providers/standards/

[6] Drummie, A and Slater, B. (2013). Master Plan vs Strategic Plan vs Business Plan. Accessed 12 August 2020. http://venues.programmanagers.com/wp-content/uploads/2016/12/2411473.pdf

IWA Publishing's authorised EU representative for General Product
Safety Regulations is Diane D'Arras, 15 rue Duret, 75116 Paris,
France, e-mail: safety@iwap.co.uk.

Printed and bound by CPI Group (UK) Ltd, Croydon, CR0 4YY

12/05/2026

02108875-0001